The Morse Code for Radio Amateurs

EIGHTH EDITION

George Benbow, G3HB

Radio Society of Great Britain

Published by the Radio Society of Great Britain, Lambda House, Cranborne Road, Potters Bar, Herts, EN6 3JE.

First published 1947
Eighth edition 1994

© Radio Society of Great Britain, 1992, 1994. All rights reserved. No part of this publication may be reproduced, stored in a retrieval system, or transmitted, in any form or by any means, electronic, mechanical, photocopying, recording or otherwise, without the prior written permission of the Radio Society of Great Britain.

ISBN 1 872309 26 7

Cover design: Geoff Korten Design.
Line illustrations: Derek Cole, Radio Society of Great Britain.
Production and typography: Ray Eckersley, Seven Stars Publishing.

Printed in Great Britain by Hillbury Press Ltd, Potters Bar, Herts.

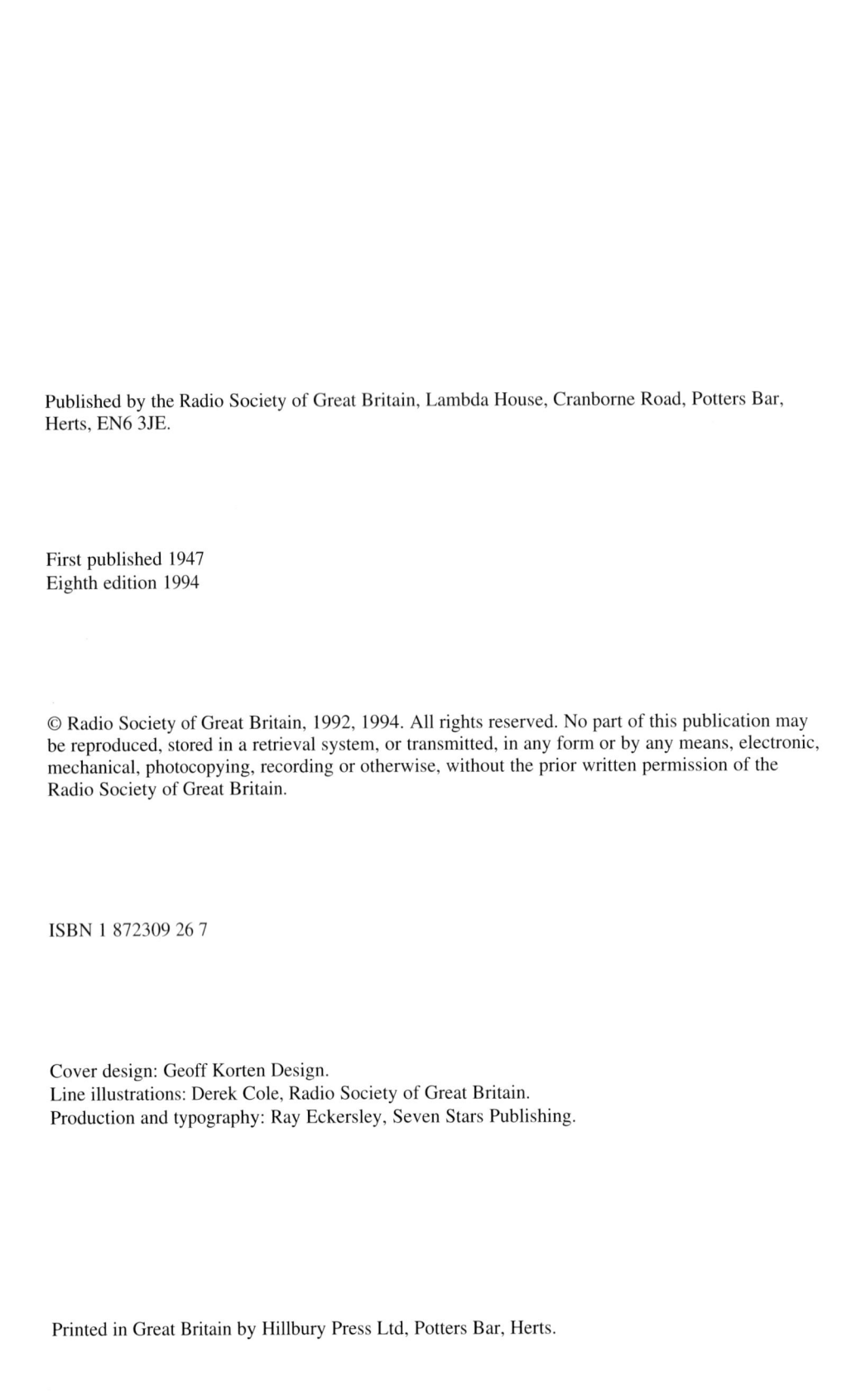

Contents

Preface	iv
1. Why Morse?	1
2. History of the Morse code	2
3. Memorising the Morse code	3
4. Reading Morse	5
5. The Morse key	7
6. Sending Morse	9
7. The Amateur Morse Tests	22
8. High-speed Morse and automatic keys	24
Appendix. Miscellaneous Morse characters	28

Preface

The Morse Code for Radio Amateurs was written in 1947 by Margaret Mills, who was one of the first members of the Women's Auxiliary Air Force to be commissioned as a Signals Officer in the Royal Air Force. She was also the first English woman to obtain a transmitting licence after the war.

It was perhaps appropriate that in 1991, the year of the 200th anniversary of the birth of Samuel Morse, this book was completely revised for the seventh edition and its scope considerably expanded, although the title and some of the original exercises were retained. This eighth edition takes into account the recent change in format of the Amateur Morse Test.

The aim is to interest people in the Morse code as a new and different language with which one can communicate worldwide, and not to regard it as an annoying obstacle in the acquisition of a Class A licence.

It will also help amateurs in self-training in communication by wireless telegraphy, a requirement of the UK licensing authority (Radiocommunications Agency of the DTI).

The book contains a brief review of the operational advantages of telegraphy, the history of the Morse code and the various types of Morse key. The main content of the book is a quite comprehensive course on the memorising, reading and sending of Morse and the various aids which are now available. This covers the requirements of the Novice Amateur Morse Test (five words per minute) and the Amateur Morse Test (12 words per minute).

Various aspects of high-speed Morse and automatic keys are discussed for those who wish to increase their speed above the latter standard test level.

It is hoped that this book will provide the answer to the question "Can I learn Morse?" – that is, "Of course you can, if you want to."

The author acknowledges with many thanks comments received from, and discussions with, Hilary Claytonsmith, G4JKS, Ray Eckersley, G4FTJ, and David Evans, G3OUF. He also thanks his wife, who typed the manuscript, having become expert in deciphering his handwriting. The photos illustrating Morse keys and equipment were kindly supplied by a number of sources and these are individually acknowledged in the captions.

George Benbow, G3HB

Chapter 1

Why Morse?

Because it's fun! But seriously, a number of advantages are gained from the use of continuous-wave telegraphy (often known as just 'CW') as a mode of radio communication, and these are outlined below.

Bandwidth

Telegraphy is the simplest form of amplitude modulation; information is transmitted by the on-off keying of the carrier wave produced by a radio transmitter to form the dots and dashes of the Morse code.

The characteristics of Morse code are communicated by the presence or absence of a 'signal'. This is commonly a signal transmitted by radio, but alternatively the 'signal' could be created by the switching on and off of a lamp (the Aldis lamp) or the reflection of the sun's rays by a tilting mirror (the heliograph) or by any other form of 'signal' which could be detected by sound or sight.

Theoretically a keying speed of about 12 words per minute occupies a bandwidth of the order of 16Hz. In practice, due to a number of factors such as the frequency stability of the transmitter and receiver, the shape of the keying waveform which may create key clicks, and deficiences in the transmitter design which cause 'thumps' and 'chirp' on the transmitted signal, a much greater receiver bandwidth is allowed.

As an indication of this, consider the bandwidths for telephony (SSB) and telegraphy which are provided in the present-day receiver or transceiver. These are typically 2400Hz for SSB and 270Hz for telegraphy. In practice a narrower bandwidth than 270Hz could be provided but at considerably greater cost.

Thus a given bandwidth can accommodate more telegraphic signals than telephonic signals, a very important point in today's crowded amateur bands!

Economics

The construction of a CW, ie a telegraphy, transmitter is relatively straightforward compared with the SSB equivalent and is well within the capability of the home constructor. (It was, in fact, mainly the advent of the complex SSB transceiver which accelerated the decline of home construction.) With careful shopping around, a 50-100W CW transmitter could be made at home, probably for less than £75. An immense amount of satisfaction can be obtained from such a transmitter used with one of the older communication receivers available on the second-hand market.

Much information is now available on the construction of low-power transmitters and simple receivers.

Operational aspects

In very weak or fading signal conditions, due to poor or changing propagation, it is very often possible to copy a Morse transmission when a telephony signal is not readable.

Similarly, CW (unlike telephony) is often readable in the presence of severe interference. The point here of course is that the much narrower bandwidth adequate for a telegraphy transmission passes far fewer unwanted signals, ie those close to the wanted signal.

The use of telegraphy generally enables much longer ranges to be achieved in two situations which are not uncommon in amateur radio. These are the increasingly popular use of very low transmitter power, ie less than 5W input, and the enforced use of antennas which are obviously not too effective, ie they are either very low or very short.

Meteor scatter and 'moon-bounce' (earth-moon-earth) communication depend on the reception of extremely weak signals, hence narrow receiver bandwidths and very high frequency stability are necessary. Morse code is used almost exclusively in both modes.

Morse and the disabled

Communication in Morse code by a person with a speech impediment is obviously possible, and quite often by those having defective hearing.

A more extreme example of the value of Morse code to a disabled person was recounted by a friend who said he felt very humble. In a contact with an American amateur, he had commented on his Morse and the American replied that he was in an iron lung and was keying his transmitter by blowing down a tube.

Chapter 2

History of the Morse code

Samuel F B Morse, son of a Congregational Minister, was born in Massachusetts, USA, in 1791 and was educated at an academy and at Yale University. Although he had studied physics, including electromagnetism and chemistry, he chose to become an artist and came to London to study painting in 1811.

He became interested in telegraphy as a result of discussions with other passengers on a voyage back to America in 1832. By the time Morse landed in America, he had acquired an almost fanatical interest in the development of a system to transmit information quickly over long distances by the electric telegraph.

He gave up painting, in which field he had achieved significant success, and his existence became inevitably that of most inventors of that age: few friends, not much encouragement and very little money.

He needed wire in very long lengths and, since this was not then available, a great deal of his time was spent in soldering short lengths together and wrapping thin thread around the wire to insulate it.

His first telegraph consisted of a pair of wires several hundred yards long with a contact at the end of each wire at the transmitting end. The contacts were made to open and close by a notched wooden rod which was drawn across them. At the far end was an electromagnet which was energised when the contacts at the transmitting end were closed. A pen attached to the electromagnet was arranged to make a mark on a tape of thin paper drawn past the pen by a spring-driven motor.

Such was Morse's first crude electric telegraph. The speed of operation was very slow due to the method of closing the contacts at the transmitting end, and also the need for there to be at each end a list of the contact closures to be made to cause the marks required at the other end, corresponding to various letters and words.

Morse made an on-off switch for the transmitting end, the 'Morse key', and published his famous code in about 1835. Interest in Morse's telegraph was not great and progress was slow. It was in May 1844 that Morse transmitted his famous message "What hath God wrought?" from Washington to Baltimore, a distance of about 30 miles.

Fig 2.1. Samuel Morse
PHOTO: SCIENCE MUSEUM, LONDON

Others now began to take an interest and many improvements to Morse's system were made. The method of 'recording' the message received on a paper tape referred to earlier was developed further and was in use for many years. The operator would transcribe the message after receipt.

It is not clear when and by whom it was realised that the incoming signal could be read by listening to the sound made by the recording mechanism. This was a major step forward and soon messages were being received and written down at speeds up to 25/30 words per minute.

Communication by the electric telegraph using Morse's code began to spread throughout America; the wires were, of course, carried on poles. The need for rapid communication during the American Civil War (1861) created further demands for the new system.

The development of land-line telegraphy in the USA in this period is very interesting, but it is not our concern here. Suffice to say that when communication by radio was first demonstrated in the last years of the nineteenth century, a means of signalling by it was available: the Morse code.

The code which Morse originated was somewhat complicated in that it contained dashes of two lengths and some letters had a space in the middle. This was known as 'American' Morse as it was used on the American land-lines but is rarely, if ever, used today.

Morse's original code was later modified: the number of dots used was reduced, letters with spaces in the middle were simplified and only one length of dash was used. This became the 'international' code, now known throughout the world simply as 'the Morse code'.

Other versions do exist. These were originated in order to deal with languages such as Greek, Russian, Arabic, Turkish etc and are virtually unknown outside the country of origin.

Chapter 3

Memorising the Morse code

Books on the Morse code mostly contain tables of the Morse characters starting thus:

A ·–   B –···   etc

This indicates that 'A' is a dot and a dash, 'B' is a dash and three dots and so on. However, Morse characters are not combinations of dots and dashes – they are 'sound patterns'. Each letter, figure and sign has its own sound pattern and it is these sound patterns and not a jumble of dots and dashes which we have to memorise. Hence the dots and dashes shown above are the only ones you will see in this book.

The sound patterns are expressed by 'dits' and 'dahs'. For ease of pronunciation 'dit' is shortened to 'di' unless it appears at the end of a character. 'Dit' or 'di' is pronounced crisply and 'dah' is pronounced as 'dar' and is slightly stressed. The Morse alphabet is given in Table 3.1.

The spacing between dits, dahs, characters and words is shown below in terms of the length of one dit.

Dah	– 3 dits
Space between elements of a character	– 1 dit
Space between characters	– 3 dits
Space between words	– 7 dits

This is shown in Fig 3.1. It is obviously vital!

There are many ways to memorise the Morse alphabet; some of the groups of letters which have been suggested are:

(a) All dits   E I S H
     All dahs   T M O
     Similars   A U V   N D B
     Opposites F L   G W   Q Y
     Inversions K R   P X
     Others   C J Z

(b) Groups of four or five letters from A to Z.

Table 3.1. The Morse alphabet

A	di-dah	N	dah-dit
B	dah-di-di-dit	O	dah-dah-dah
C	dah-di-dah-dit	P	di-dah-dah-dit
D	dah-di-dit	Q	dah-dah-di-dah
E	dit	R	di-dah-dit
F	di-di-dah-dit	S	di-di-dit
G	dah-dah-dit	T	dah
H	di-di-di-dit	U	di-di-dah
I	di-dit	V	di-di-di-dah
J	di-dah-dah-dah	W	di-dah-dah
K	dah-di-dah	X	dah-di-di-dah
L	di-dah-di-dit	Y	dah-di-dah-dah
M	dah-dah	Z	dah-dah-di-dit

(c) Groups based on the frequency at which letters are used, such as:

AEDST BILNR DFGU CKMP HQWY JVXZ

No doubt other combinations have been suggested.

A possible disadvantage of the above is that, in order to remember a particular letter, you have first of all to recall the group in which you learnt it. So why not start at A and finish at Z? You may find it easier!

Some basic rules for learning are:

(a) Never study when tired.
(b) Study periods should be short, say 20 minutes.
(c) If at all possible, study should be regular. Two short periods per day are much more useful than an occasional two hours.

Whatever method you use to learn the letters, concentrate on a few at a time and only learn more when you have mastered the first group.

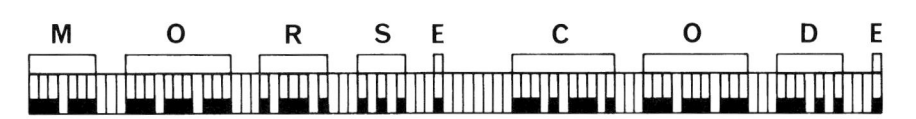

Fig 3.1. Spacing diagram for Morse code

Table 3.2. Figures in Morse

1	di-dah-dah-dah-dah
2	di-di-dah-dah-dah
3	di-di-di-dah-dah
4	di-di-di-di-dah
5	di-di-di-di-dit
6	dah-di-di-di-dit
7	dah-dah-di-di-dit
8	dah-dah-dah-di-dit
9	dah-dah-dah-dah-dit
0	dah-dah-dah-dah-dah

So now you can start to memorise the Morse characters. It does not matter how long this takes – half an hour or a week.

The next step is familiarisation. When you see or think of the letter 'B', you must instinctively call up from your memory the Morse character dah-di-di-dit. This is really the most important part of learning the code. You can practise anywhere; translating car registration letters (if you see 'ULF', say to yourself "di-di-dah, di-dah-di-dit, di-di-dah-dit"), advertisements in the train or on hoardings, newspaper headlines and so on. It is better to 'dit and dah' away to yourself and not aloud, otherwise people around you might wonder!

Rest your right (or left) hand on your knee or chair arm etc and make a slight up and down movement of your thumb and first two fingers in sympathy with your dits and dahs (but not when driving, please!) The object is to simulate the pressure you will be putting on the knob of a Morse key. This movement must also become instinctive.

Now for a change, what about the figures from 1 to 0? The Morse characters for these are shown in Table 3.2, and are easier as the pattern is obvious.

The procedure and punctuation signals shown in Tables 3.3 and 3.4 are widely used in amateur radio and are required for the tests. The bar over certain signals in Table

Table 3.3. Procedure signals

End of message ($\overline{AR}$)	di-dah-di-dah-dit
End of work ($\overline{VA}$)	di-di-di-dah-di-dah
Wait ($\overline{AS}$)	di-dah-di-di-dit
Invitation to transmit (general) (K)	dah-di-dah
Invitation to transmit (specific station) ($\overline{KN}$)	dah-di-dah-dah-dit
Received all sent (R)	di-dah-dit
Station closing (CL)	dah-di-dah-dit di-dah-di-dit

Table 3.4. Punctuation

Stroke (/)	dah-di-di-dah-dit
Break sign (=)	dah-di-di-di-dah
Hyphen (-)	dah-di-di-di-di-dah
Question mark	di-di-dah-dah-di-dit
Full stop	di-dah-di-dah-di-dah
Comma*	dah-dah-di-di-dah-dah
Erase	di-di-di-di-di-di-dit

* Also used as exclamation mark.

3.3 denotes that the characters are sent as one, eg $\overline{AR}$ is sent without a space between the 'A' and the 'R'.

Chapter 11 of the *RAE Manual* contains a comprehensive explanation of the use of the above and other signals. The procedures which are typically used in amateur telegraphic communication are also explained in detail.

The familiarisation process defined above must continue until you are confident that you can convert letters and figures into sound patterns almost instantaneously, and at the same time make the corresponding movement of the thumb and first two fingers to simulate the action of keying. Then we can move on to a more interesting stage and find a source of well-sent Morse to listen to.

Chapter 4

Reading Morse

Let us assume that you have now familiarised yourself pretty well with the sound pattern of each Morse character. It may have taken you longer than expected, but do not be discouraged! The time has now come to start the real thing, taking down Morse (some will say "copying Morse" or "reading Morse").

The speed of Morse code is measured in 'words per minute' (wpm) and the average length of a word is taken as five letters. There is no 'standard' word, eg 'books' is longer than, say, 'seeds' in Morse; but words such as 'Morse' and 'Paris' have been suggested. In speed competitions, speed may be measured in 'characters', ie 150 characters is taken as 30wpm.

A number of automatic Morse code senders are now available. These will produce random groups of five letters or five figures and the sending speed can be adjusted from about 5wpm to about 36wpm. All spacing with reference to the dit length is correct at all speeds. If you are able to record the output of the sender, you can then check what you have copied.

Morse for practice purposes is also available recorded on cassette for use in the standard domestic tape recorder. The following tapes are available from RSGB HQ:

Stage 1	5wpm
Set 1	5-10wpm
Set 2	10-15wpm
Set 3	15-22wpm

Sets 1-3 contain two 90 minute cassettes having words, sentences, random code groups, related text and practice contacts. The individual Morse characters are recorded at 18wpm, but the spacing between characters is adjusted to give you more time initially to read the letters and numbers. Tapes and records produce perfect electronically controlled Morse code and listening to these will also give you a feeling of what to strive for in your sending.

If you have a computer at home, a third attractive possibility is to use this in conjunction with a suitable program. These are commercially available for most computers (the PSION organiser has its own Morse tutor); they can send Morse code to help when learning, and also later when a large amount of practice is required. No computer skills are required, and no extra equipment is necessary. Typical commercial programs can send Morse from 5wpm upwards to 30wpm or more, in random groups (restricted to certain letters at first) or as practice texts. If you can program in BASIC you should be able to write your own without too much difficulty – a simple Morse practice program is included in the book *Amateur Radio Software*, published by the RSGB. However, ready-made programs are inexpensive and most people wish to get on with learning Morse rather than learning programming.

Slow morse transmissions are organised by the RSGB. These are from a number of amateurs throughout the country and are on the amateur 1.8MHz and 144MHz bands. Current details are available from the Membership Services Department, RSGB Headquarters.

Some radio clubs organise Morse code classes.

Two other Morse transmissions at higher speeds which will perhaps be of interest to you later on are as follows:

(a) The RSGB News Bulletin is now transmitted at 12-22wpm (current details from MSD at RSGB HQ.)
(b) The Headquarters Station of the Royal Navy Amateur Radio Society, G3BZU, transmits a code test at 15-40wpm currently on the first Tuesday of each month at 2000hrs clock time on 3.600MHz.

It is advisable at this point in your learning curve not to listen on the amateur bands. Some of the Morse heard here is not particularly good. At this stage you are unlikely to be able to distinguish between very good and not-so-good Morse. You must not pick up bad sending habits at this stage!

It should perhaps be mentioned again that the individual Morse sound patterns you have been listening to from tapes etc are sent at a speed of 16-18wpm. This is what they sound like in use, ie not a dah-di-di-dit which lasts two or three seconds.

Morse is generally taken down in longhand, so think about your handwriting; it must be clear and legible without any fancy flourishes. Do you need to try to change your style? How quickly can you write legibly? – probably not much more than 20-25wpm. A professional operator will copy at up to 35 or 40wpm, but not for long periods.

You will probably find it easier to use block capitals at

5wpm. *You must fight this temptation and use longhand from the very beginning.* As your speed increases you must not have to face up to changing from block capitals.

Pen, ballpoint or pencil? A fairly soft pencil (HB or B) is generally to be preferred and is safer than running the risk of an empty pen or messy ballpoint for the test.

You may start listening to the 5wpm tape and later to the 10wpm one. Remember that each character is sent quite fast, about 18wpm, but the spacing between characters is adjusted accordingly.

It is important to write each letter down as you receive it. If you are not sure what it was, ignore it and leave a space, otherwise you may lose the next two or three letters.

You may well find that progress is somewhat erratic – many learners hit a plateau around 10wpm and do not seem able to progress further. If this happens, drop all practice for a week or so and you will probably find that you can then read a higher speed than before.

You may have access to a random Morse sender and do not forget the RSGB Slow Morse Transmissions.

The disadvantage of Morse tapes is that after a while you begin to remember what is coming next. However, with tapes and/or an automatic sender you will spend a considerable time listening to correctly formed Morse code – this will be very valuable preparation for when you start sending it. Generally quite a number of commercial code transmissions can be heard on a communication receiver. As you get familiar with the sound of Morse, start listening on the amateur bands – the 7MHz and 3.5MHz bands are likely to be of most use. Class B licensees can now use Morse on the air for practice purposes. The guidance of a local licensed amateur or an amateur radio club is invaluable at this stage.

The code speed on the amateur bands varies from around 10wpm to about 25wpm. The average is probably about 16 to 18wpm. In the Novice sections of the bands the speed is around 5 to 12wpm.

The old adage "practice makes perfect" is really true. Listening to good Morse code and trying to emulate the rhythm is by far the best way to learn Morse and build up the reflex action.

Chapter 5

The Morse key

There is no such thing as a standard Morse key. Literally thousands of different patterns have appeared since Morse made the first one in 1835.

The vast majority of keys fit into two categories: the first type has a fairly substantial horizontal arm, generally made of brass, pivoted towards one end. The knob is at the front end and the moving contact is on the underside of the arm between the pivot and the knob. Behind the pivot is a tension spring and an adjustable stop to limit the angular movement of the arm and thus the gap between the contacts. Basic details are shown in Fig 5.1.

The knob may be up to 60mm or more above the table. In many of the older keys of this type, the back stop may incorporate contacts, hence the key has 'back' contacts as well as main contacts. The back contacts open before the main contacts and could be used in the send/receive switching function. Many British keys fit into this category.

Fig 5.2 shows a typical key of this type. The under surface of the wooden base is stamped 'GPO 6151', and the key is believed to have been made by Marconi Co.

The other pattern of key, often American in origin, is smaller and has a lighter arm which is cranked downwards at the knob end. Hence the knob is only about 20mm or so above the table. This type of key is shown in Fig 5.3 and this particular one was made by McElroy (USA). It cost the equivalent of 35 pence in 1938 and is still in use.

Keys of this type are known as 'straight' keys and sometimes as 'up and down' keys. All keys, irrespective of type, will have adjustment for the gap between the main contacts and the tension of the spring which holds the contacts apart. Older keys often have a somewhat crude switch on the base to short the contacts if required.

Fig 5.2. GPO straight key

An interesting development in the modern key is the use of small ballraces in the pivot of the arm. Although the angular movement of the arm is very small, the 'feel' of the key in use is very much improved.

Morse keys are available from a number of sources. Imported keys may be obtained at prices up to £100 which seems somewhat expensive in amateur radio. British-made keys incorporating ballraces are now obtainable at quite reasonable prices. A modern key of this pattern is shown in Fig 5.4.

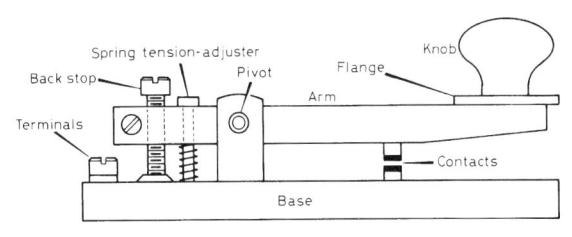

Fig 5.1. Outline drawing of typical morse key

Fig 5.3. American straight key (circa 1938)

Fig 5.4. A modern straight key

PHOTO: R A KENT (ENGINEERS)

Occasionally ex-Service keys appear on the surplus market. Second-hand keys are often somewhat old, but nevertheless very good. However, they are classed as 'collectors' items' and carry an appropriately inflated price.

Toy keys sold as part of cheap Morse practice sets are often worse than useless.

Mounting and holding the key

The position of the key on the operating table is very important. The larger pattern of key was intended to be mounted at the edge of the table, and thus the forearm is more or less horizontal and unsupported whilst keying. The smaller key, having a lower knob, is most conveniently mounted towards the rear of the table so that the forearm and wrist rest on the table, which is a less tiring position. If the larger key is used in this position, the wrist is often bent upwards at quite a sharp angle.

The standard way to hold the knob is with the thumb on one side, the tip of the first finger on the top and the tip of the second finger on the other side. The exact positions depend to some extent on the shape of the key knob and whether it has a flange. The key should be held firmly, but not tensely.

A comfortable operating position is essential for long sessions!

The contact gap and spring tension

To start with, the gap between the main contacts should be such that the vertical movement of the knob is about 1½ to 2mm. Spring tension should not be too loose; you must not produce stray and unwanted dits.

As you gain experience, you will probably find that you want to reduce the contact gap and loosen the spring.

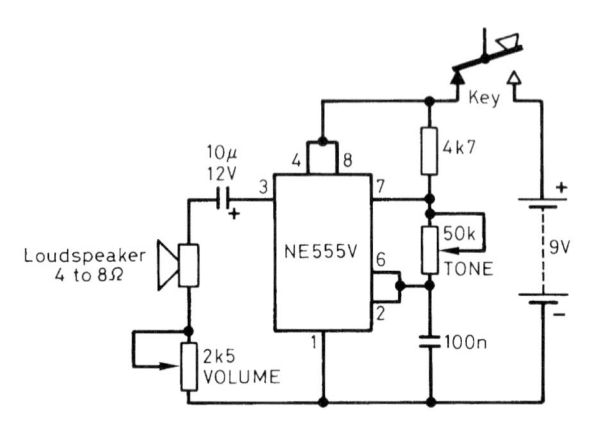

Fig 5.5. Circuit of Morse practice oscillator designed by WA5YFL

Practice oscillators

The simplest arrangement is the key wired in series with a buzzer and a battery. This is not ideal as the note produced is very low, quite unlike the note from a radio receiver. A simple, widely used, electronic oscillator originated by an American amateur is shown in Fig 5.5. It can be easily be made using Veroboard circuit board.

Chapter 6

Sending Morse

In the last chapter we discussed Morse keys, how to adjust them and how to use them. You may have been surprised that this was the first real mention of the key. This is intentional.

By now, if you have followed the order of this book, you will have listened to a lot of good Morse from tapes etc and started to listen on the amateur bands. You will have reached the 5wpm receiving requirement of the Novice Amateur Morse Test, and you are possibly well on the way to meeting the 12wpm requirement of the Amateur Morse Test.

It is time, therefore, to start sending Morse and you will now be imitating the sound patterns to which you have listened for so long. Thus it will be much easier for you to send properly formed Morse and to recognise your own mistakes.

So, adjust the contact gap, spring tension and position of your key. Make sure that you are comfortable. Don't forget that in the tapes etc you have heard, the individual letters are sent fairly fast, equivalent to 16-18wpm, and it is the adjustment of the spaces between letters which gives the timed words per minute. You must make the same adjustments. Now start Exercise 1. When you feel you have mastered this one, go on to the next, periodically revising the previous exercises. If it is at all possible you should, occasionally at least, send to someone who is experienced in Morse. This is the best way to get your spacing correct.

EXERCISE 1 – E, I, S, H, T, M, O

E I S H E	T M O T M	E T I T O	H O S T S
I H E S I	T O M O T	S I M S M	H O I S T
S I E S H	M O T M O	T E S O H	T H O S E
H S S E I	T M O T M	O M I H E	S H O T S
I E H S E	O M O T O	T M O S E	T H E S E

The following words may be practised in any sequence.

TO	TOM	MISS	SHOT	THEME
IS	HOT	HISS	MIST	THESE
HE	SHE	TOME	TEEM	THOSE
IT	THE	SOME	HOSE	TOTEM
SO	HIT	THEM	HOOT	MOIST
	TIM	HIST	HOST	
	TOE	MOST	MITE	

TIMED WORDS

Sent in 1 minute = 12 words per minute; in 2 minutes = 6 words per minute

MIST	MITE	THEM	SOME	THESE
MOST	TOE	HIM	MISS	THOSE
SHOT	TOM	TOTE	HOST	TEEM

THOSE	THESE	SHOTS	HOSTS
TOTEM	HOIST	MITES	MOIST
HOMES	THEME	MISTS	HOSES

EXERCISE 2 – A, U, V

E O M U T	A V A A U	A T E M U
E I S O V	V U A V A	E U I A S
M I S U A	A U U A V	V H V T M
S O S T V	V U V V U	O U E A I
U H I T O	I T U I T	V S A H U
S T V S T	S T V T S	T A M V O

AT	HAT	THAT	SHUT	MAVIS	STOVE
TO	EVE	MEET	VOTE	MAUVE	STOUT
HE	SAT	HIVE	MOAT	HOUSE	STAVE
IS	MAT	OUST	MATE	MOUSE	SHAVE
IT	HUT	HUSH	HATE	STEAM	SHAME
US	EVA	MESH	TAME	SUAVE	VISIT
AS	USE	SHAM	VAST	TEETH	SEAMS
ME	OUT	SOOT	MUSE	TITHE	SHOUT
SO	VET	HAVE	MUST	EAVES	SHEAVE
AH	VIE	SAVE	THOU	SHEET	MOUSSE

Sending Morse 11

TIMED WORDS

Sent in 1 minute = 12 words per minute; in 2 minutes = 6 words per minute

THAT	HAVE	THESE
MATS	SAME	THOSE
SEAM	VOTE	SHOUT
SHUT	MESH	MOUSE
MATE	SOOT	

Sent in 1 minute = 12 words per minute; in 2 minutes = 6 words per minute

HOUSE	MAUVE	SHAME
MATES	SHOUT	TEETH
STEAM	EAVES	STOUT
STAVE	TITHE	MAVIS

Sent in 3 minutes = 12 words per minute; in 6 minutes = 6 words per minute

AS	HAT	SHAM	MUST	STOVE	STAVE
US	SAT	THAT	HUSH	STOUT	SHAVE
AT	MAT	SAVE	MUSE	SUAVE	TITHE
	EVA	OUST	MESH	TEETH	EAVES
	HUT	HAVE	SOOT	SHAME	SEAMS
	USE	SHUT	METE	VISIT	MAVIS
	VET	VOTE	TAME	MAUVE	MOUSE
		MOAT	VAST	HOUSE	STEAM
			SEAM	SHOUT	

EXERCISE 3 – N, D, B

N V A N B	N S O V N	V S B I V
A D N U B	E D H A D	D U E N A
D U B D V	T S B T S	V N U O D
U B V A B	A M S B M	I T D I T
N V A N U	V U O T H	U S H D U
B A B V U	N T V B E	I H E A I
N U D A N	N A E H I	E B V T N
A N B V A	U D S H U	H T O U V
V D B V A	T I D T I	M B S M A
N U D A V	D O U N V	S T B S T
N D B V A	A N E U D	D A H E D
A B D V A	V I B S U	N V O S N

ON	ODD	BON	BUST	BAND	BENT	HOUND	ABBOT	AVENUE
DO	ADD	BUS	TONE	TIME	DONE	ABOUT	VOTED	TOMATO
BE	ODE	ANT	NOTE	DAVE	BONE	SOUND	BOAST	BOVINE
IN	AND	MOB	DOME	SENT	BOMB	ABOVE	DAUNT	BOTTOM
AN	TEA	NOT	DOTE	HAND	DENT	BEAST	BENDS	VETOED
	TON		BOAT	BEST	BASS	MOUND	NOTED	VESTED
			BOSS	BEAM				MOBBED
								TEASED

TIMED WORDS

Sent in ½ minute = 12 words per minute; in 1 minute = 6 words per minute

ON	DO	BE	IN	AN	AT	TO	HE	IS	IT	US	AS	SO	ME	ON

Sent in 1 minute = 12 words per minute; in 2 minutes = 6 words per minute

AND	THE	TEA	TON	BOB	NOT	ANT	BUS	ODD	SAT
EVE	HAT	TOM	SHE	HIT	HIM	HOT	TIM	MET	SAD

Sent in 1 minute = 12 words per minute; in 2 minutes = 6 words per minute

HOUND	BENDS	ABOUT	DAUNT	SOUND	VOTED
ABOVE	BEAST	ABBOT	BOAST	MOUND	NOTED

Sent in 3 minutes = 12 words per minute; in 2 minutes = 6 words per minute

AVENUE	TOMATO	BOVINE	TEASED	MOBBED	VETOED	BOTTOM	VESTED
ABBOT	BOAST	VOTED	NOTED	HOUND	MOUND	ABOUT	
SOUND	ABOVE	BEND	DENT	BENT	BOMB	DONE	BONE
BAND	TIME	DANE	SENT	BEST	HAND	BASS	BUST
BEAM	TONE	BOSS	BOAT	NOTE	DOME	DOTE	BOB

EXERCISE 4 – A, W, J

A N D W A	A N U E T	M E M E U
J W D J B	E W D V I	D J W T M
B W J B N	M I J B U	M D H O V
J A N A A	J M S A N	J E M U D
D W N D W	O H V D W	N W B A T
W A B N A	E U B J S	B A I M H
D J D A B	V N A T V	V B A N S
W D J N A	D W I M H	M J I A V
N A W N W	J B U S O	I M H T U
A D J B D	E A N V H	T E U N A
J N A E D	M T W J D	J M U S O
B D A J D	E U M E M	D V E O J

WE	HEW	WENT	JOVE	JOAN	JAUNT	DOBBIN
WIT	VOW	WISH	JANE	JUDO	JOUST	WANTED
JOE	VIE	WANT	JADE	WAVE	SHOWN	WISHED
JOB	JUT	WEST	JIVE	VANE	WHIST	WEEDED
JOT	OWN	JEST	MOWN	VAIN	WEEDS	SEEDED
JET	OWE	JOIN	TOWN	DUET	SWEDE	JESTED
WET	TOW	WHAT	VOWS	SUET	JUMBO	HAUNTED
VET		HEWN	JUNE	WHEN	OWNED	
		SEWN	JUTE	JOBS	TOWED	
		SOWN			WHOSE	
					JOINT	

TIMED WORDS

Sent in 1 minute = 12 words per minute; in 2 minutes = 6 words per minute

DUET	JANE	TOWN	WAVE	VANE	WHAT	WENT	JOIN
SOWN	WEST	WHEN	JUNE	JOB	OWN	USE	VET

JOINT	OWNED	TOWED	WHIST	SHOWN	SWEDE
WHOSE	JAUNT	WEEDS	JOINS	WANTS	JUMBO

Sent in 3 minutes = 12 words per minute; in 6 minutes = 6 words per minute

TOW	WIT	OWE	JOE	OWN	JOB
JUT	JOT	VIE	JET	VOW	WET
HEW	VOW	JIVE	WENT	JADE	WISH
WANT	JANE	WEST	JOVE	JOIN	SOWN
WHAT	SEWN	HEWN	MOWN	JOB	TOWN
WHEN	VOWS	SUET	JUNE	DUET	JUTE
VAIN	JOAN	VANE	JUDO	WHIST	JAUNT
SWEDE	JOUST	WEED	SHOWN		JESTED

EXERCISE 5 – C, K, P, G

C A J W P	C E I S C	O M G T G
J K A K A	H K T K M	O H J C V
K J P W P	O A P U V	C S O C W
A G J G W	A G J G W	V M V C J
G A C J G	C N D B C	C H O A C
P W J A C	E I K I S	E K A G N
K A K J A	H T P M O	K T G N K
P W J W P	A G U G V	S K U G B
W G J G W	C N D B C	K M G B K
G A K A J	A K W K J	G S K B M
A K J G A	E I P S H	K N G T K
P A J W K	T G M G O	E K G N T

DOG	APT	PACK	BAKE	PATH	COAT	PASTE	GHOST
PAT	GOT	TACK	SAKE	PAWN	GOWN	DUTCH	WAGES
CAT	GET	JACK	WAKE	DUCK	JOKE	AWAKE	PAGES
ACT	COT	KATE	AGED	MUCH	WAGE	GUESS	GAUGE
APE	AGO	CAKE	GOAT	COMB	PAGE	GUEST	GAINS
AGE	TAG	CAPE	POKE	COVE	CUTE	COAST	AGAIN

TIMED WORDS

Sent in 1 minute = 12 words per minute; in 2 minutes = 6 words per minute

PACK	PAGE	TACK	WAGE	MUCH
CAKE	PAWN	COMB	SAKE	DUCK
JACK	COVE	GOAT	GOWN	COAT

PASTE	DUTCH	AGAIN	GAUGE	AWAKE	GAINS
PAGES	GUESS	WAGES	GHOST	COAST	GUEST

Sent in 3 minutes = 12 words per minute; in 6 minutes = 6 words per minute

DOG	PACK	PATH	PASTE	NAG	TACK	PAWN	DUTCH	JACK
DUCK	AWAKE	PAT	KATE	MUCH	GUESS	CAT	CAKE	COMB
GUEST	ACT	BAKE	COVE	COAST	APE	SAKE	COAT	GHOST
AGE	WAKE	GOWN	WAGES	APT	AGED	JOKE	PAGES	GOT
COAT	WAGE	GAUGE	GET	PAGE	GAINS	COT	AGAIN	AGO

EXERCISE 6 – R, L, Q, Z

C K P G R	Q A R E L	W P C V L
L Q Z K P	C Z M U D	J G R O Q
Z G K P Q	J K P G Z	N M Z H K
L P K G L	T Q V B L	D T C G J
R G K C P	W S K C J	B O K Z B
P L Q C K	G H P M T	W J B N D
Z K L C Q	B D N J W	G Q O R G
P Z K C K	O T M G P	B Z K O B
P K Z K C	K R Z R C	C J H Z W
L C Q G R	Z G H O V	K C Z C R
Q C Z R L	B J K Q L	G W T J C
Q L P Z C	R C P G K	L E Q R L

LONG	RITZ	ZEPP	QUEEN	REQUIRE	PRESENT
LAST	RATE	ZEAL	QUITE	REQUEST	PROMISE
LOST	ROTA	ZERO	QUEST	RELATE	PROGRESS
LATE	REAL	ZONE	QUIET	REGALE	PROPER
LOOM	ROAN	ZOO	QUOTE	RESET	PROSPECT
LIKE	RAVE	ZEBRA	QUEER	REPEAL	PROGRAMME
LOOK	ROLL	ZOOM	QUIRE	REPEAT	PRESS
LINE	RING	ZIP	QUAIL	REPENT	PRESET
LEAP	REST		QUILT	REPAST	PROVE

Sent in 1 minute = 12 words per minute; in 2 minutes = 6 words per minute

| PRESENT | RITZ | QUEEN | LEAP | QUEER | PROVE | REST |
| ZEBRA | QUEST | REPAST | LONG | ZONE | WE | |

| PROGRESS | ROAN | ZEAL | QUIET | REQUIRE | RAVE |
| LOOK | ROLL | REPEAL | ZEPP | QUOTE | QUEEN |

Sent in 4 minutes = 12 words per minute; in 8 minutes = 6 words per minute

PRESENT	REQUIRE	PROMISE	REQUEST	QUEEN	QUITE	PROGRESS
RELATE	QUEST	PROPER	REGALE	QUIET	PROSPECT	RESETS
QUOTE	PROGRAMME	REPEAT	QUEER	PRESS	REPEAL	LAST
REPENT	REPROVE	REPAST	QUEEN	ZEPP	RITZ	QUITE
ZEAL	RATE	QUEST	ZERO	ROTA	QUIET	
REAL	QUOTE	ZOO	ROAN	QUEER	RAVED	
ROLL	RING	REST	LIKE	ZONE	ZEBRA	

EXERCISE 7 – F, X, Y

R L Q Z X	A Z Q J I	Q W E R T
Q Z F X Y	G B Y P H	Y U I O P
Y X F Z Q	A F C X O	A S D F G
L R Z Q Y	H B E D W	H J K L Z
R Q Y L R	J I C N E	X C V B N
Z X Q L R	K G M F V	M P O I U
Y X F Z Q	N L G U B	Y T R E W
Z X Q L R	K H T B A	Q L K J H
L R Q Y Z	I S D C M	G F D S A
X Y Z Q L	R O F R E	M N B V C
R L F Z Q	T U Q P L	X Z T Y U
Y F X Z Q	C A V Z S	Q P A L Z

YAGI	YACHT	FIRST	FATE	LAZY
YEAR	YEAST	FAULT	FENCE	HAZY
YARD	YIELD	FAST	FERRY	CRAZY
YARN	YOURS	FEAST	FIGURE	MERRY
YAWL	YOUTH	FLIGHT	PIXEY	BERRY
YAWN	YEOMAN	FISH	EXPIRE	TERRY
EXIT	MIXTURE	WAXEN	EXPLODE	EXHIBITION
VIXEN	TEXTURE	EXCLAIM	EXPRESS	HEXAGON

TIMED PASSAGES

Each passage sent in 3 minutes = 14 words per minute; in 7 minutes = 6 words per minute

When preparing the screened wiring it is essential to avoid leaving frayed ends of the screening braid which could cause trouble A neat way of finishing the braid is to bind each end for about three eighths of an inch with a close winding of number twenty two S W G tinned

copper wire a tail of this wire being left with which to earth the braid to the nearest earth point Any attempt to solder the braid directly to an earth tag will often result in melting the P V C covering with a consequent short circuit Where several wires are run

together they should be bunched and bound with a few turns of twenty two S W G wire at convenient points For general wiring number twenty two S W G P V C covered wire should be used the screening braid being cut from screened wire usually multi stranded

and thus unsuitable for general wiring The braid cover is concertinaed to enable it to slide off the original wire and on to the prepared length of P V C covered wire which should protrude for about one inch at each end Heater wiring should of course be made in heavier

Each passage sent in 3 minutes = 15 words per minute; in 7½ minutes = 6 words per minute

> In order to obtain the best keying characteristic it is recommended that the cathode of V3 should be keyed On standby switch SW3 may be used or a remote control switch or relay plugged into the oscillator control jack The usual tests should be made for the stability of the PA using

> a dummy load comprising a combination of resistors or a lamp load having an effective total resistance of 80 ohms and capable of dissipating some 30 watts of RF This is connected by short leads to the coaxial output socket Providing the design has been faithfully followed there

> It should be remembered that the coupling to the aerial tuning unit should be reduced slightly from the optimum CW condition to ensure upwards movement of the feeder ammeters on modulation ie incremental modulation It is essential to check that the heater voltage measured at the

EXERCISE 8 – SPECIAL PRACTICE SENDING GROUPS

It will be noted that all these groups, if sent badly, could be read to have more than one meaning; for example 'MIZMI' could be read as 'ZZZ' if the letters are incorrectly formed. For this reason great attention should be given to sending them.

M I Z M I	A I L A I	A O I A O	O I 8 I O
N M Y N M	I T U I T	W M I W M	M D 8 M D
D T X D T	A N P A N	J T I J T	M G 9 M G
S T V S T	T U X T U	I O 2 I O	O N 9 N O
E I S I E	A E R A E	U M 2 U M	M O 0 M O
E O J E O	T A K T A	S M 3 S M	9 O N M G
N N C N N	W E P W E	V T 3 V T	8 M D O I
T I D T I	U E F U E	H T 4 H T	7 M S T B
T S B T S	E M W E M	M S 7 M S	6 T H B E
M E G M E	T K Q T K	T B 7 T B	4 H T E V
G T Q G T	T W Y T W	G E 7 G E	3 S M V T
T E N T E	K E C K E	9 M O O M	2 U M I O

ZMITD	YTWNM	YKTTW	NTETA
YNMTW	QGTTK	CKENN	LAIED
XDTTU	XTUDT	LEDAI	UITEA
CNNKE	BNITS	ZDTMI	PANEG
VSTIA	GTNME	MEGTN	XNADT
SEIIE	KTANT	FUIEN	RAEEN
JEOAM	PWEAN	BTSNI	VIAST
DTINE	FINUE	GMETN	JAMEO
QGTMA	WATEM	WEMAT	PEGAN

EXERCISE 9 – PRACTICE GROUPS OF MIXED LETTERS AND FIGURES

63CDE	FY123	A412B	P7L34
35R9Y	Q9P32	AX2L5	6BMN3
1Z7MN	ET7QE	FY736	VU2M8
I31A2	XC3IB	VL6ZT	5TEHS
ND8B3	JAW13	6RMY3	KGM8W
BK097	DT7U2	P23Z7	S91HS
7A4FG	8L7Z3	1L9VU	K9473
HKO97	4VB6C	8735H	L3N12
PRJY3	50R3G	FSY75	QYZFL
LTB79	31NO7	E73 1I	42P8L

EXERCISE 10 – TYPICAL AMATEUR RADIO CONTACT MESSAGES

These are based on those used in the Morse Tests. If sent in the approximate times quoted, ie 6 minutes or 2½ minutes, or at 2½ or 1½ minutes, the speeds would be 5 words per minute or 12 words per minute respectively.

SENDING

NZ3CPK DE G9AVF = GM RALPH UR RST 479 WID QRM NAME IS JIM ES QTH IS WREXHAM OK? NZ3CPK DE G9AVF KN

SM5XYZ DE G7AAA = GE FINN TNX CALL UR RST 589 NAME IS TOM QTH PORTSMOUTH OK? AR SM5XYZ DE G7AAA KN

LZ9QX DE G7AAA = R MNI TNX FER FB QSO QSL OK HPE CUAGN 73 GB AR LZ9QX DE G7AAA VA CL

G0OO DE VK2XYZ = R OK UR RST 459 IN SYDNEY WAT IS UR POWER ALSO ANT? TEMP HR IS 28C AR G0OO DE VK2XYZ KN

UM7QK DE GM9ABC = GM VLAD TNX FER RPRT UR RST 459 QTH IS LONDON ES MY NAME IS IS GEORGE UR QTH AGN PSE HR POWER IS 125 WATTS QRU? AR UM7QK DE GM9ABC KN

VK6AB DE GW3ZYL = GM TOM UR SIG 589 FB HR IN OXFORD TNX FER QSO IN BERU CONTEST NW USING G5RV ANT = HW IS WX IN PERTH? AR VK6AB DE GW3ZYL KN

RECEIVING

3F2KZD DE F7XVQ = GE OM MNI TNX FER CALL UR RST 589 WID QSB = QTH IS 19 KM NORTH OF PARIS NAME IS PIERRE = RIG IS TS 530S ANT IS DIPOLE HW CPY? AR 3F2KZD DE F7XVQ KN

G9OKL DE HB9QP/P = R TU RPRT OP IS FRITZ QTH NR ZURICH WX IS SUNNY ES TEMP IS 24 C = CAMPING HR WID 40 SCOUTS FR 3 WEEKS HW CPY? AR G9OKL DE HB9QP/P KN

G9XYZ DE WQ4CKL = GE OM ES MNI TNX FER UR CALL VY PSD TO QSO UR RST 469 MY NAME IS ELMER QTH IS 40 MILES SOUTH OF DODGE CITY WX TODAY IS SUNNY ES TEMP IS 20 C OK? AR G9XYZ DE WQ4CKL KN

VK2XYZ DE G7OO = R OK RIG IS TS530S ES ANT IS HALF SIZE G5RV ABT 16 FEET HIGH WX FINE BUT COOL PSE QSL CONDX HR NOT GUD SO WILL QRT SOON QRU? HPE CUAGN SN VY 73 GB ES GL AR VK2XYZ DE G7OO KN

JA2YXZ DE AK0LB = GD OM TU FER RPRT UR SIG RST 45/79 QTH IS COLORADO MY NAME IS CHUCK USING KW TO 5 ELEMENT YAGI WX IS COLD WID SNOW ON GROUND QRT SN QRU? AR JA2YXZ DE AK0LB KN

ZL4ABL DE G7XYT = GD OM MNI TNX FER CALL UR RST 46/79 BAD QRM ES SUM QSB NAME IS JACK USING TS830 ES 3 ELEMENT BEAM PWR IS 110 WATTS WX HR IS COLD AND WINDY THIS IS MY FIRST QSO WID ZL SO SURE WUD LIKE UR QSL PSE TNX QSO HPE CUAGN 73 ES DX GB AR ZL4ABC DE G7XYT VA

Chapter 7

The Amateur Morse Tests

International regulations require that those who operate on the HF bands which have the potential for long-distance communication must have a knowledge of the Morse code. The necessary Morse tests are now administered by the Radio Society of Great Britain on behalf of the Radiocommunications Agency (the licensing authority for the UK). A pass in these tests is valid for life.

The tests are in the format of a typical contact between two radio amateurs. Thus candidates have to be familiar with some of the common abbreviations, Q-codes and procedural/punctuation characters used in amateur communication, and must be able to send and receive mixed letters and figures.

UK radio amateurs will therefore be familiar with telegraphic communication before they receive their licences and so should be much more confident in the use of a new language in which they can communicate world-wide.

The tests can include any of the following commonly used abbreviations, Q-codes and procedural/punctuation characters:

Abbreviations

ABT	–	about	PSE	–	please
AGN	–	again	PWR	–	power
PWR	–	power	R	–	roger
ANT	–	antenna	RPRT	–	report
CPI	–	copy	RST	–	readability, signal strength, tone
CPY	–	copy			
CUL	–	see you later	RX	–	receiver
DE	–	from	SIG	–	signal
ES	–	and	SRI	–	sorry
FB	–	fine business	TEMP	–	temperature
FER	–	for	TKS	–	thanks
FM	–	freq mod	TNX	–	thanks
GA	–	good afternoon	TU	–	thank you
GD	–	good day	TX	–	transmitter
GE	–	good evening	TCVR	–	transceiver
GM	–	good morning	RX	–	receiver
HPE	–	hope	VERT	–	vertical
HR	–	here	VY	–	very
HVE	–	have	WID	–	with
HW	–	how	WX	–	weather
MNI	–	many	XYL	–	wife
MSG	–	message	YL	–	young lady
NW	–	now	73	–	best regards
OC	–	old chap	88	–	love and kisses
OM	–	old man			

Q-codes

QRA QRG QRK QRL QRM QRO QRP QRQ QRS QRT QRV QRX QRZ QSA QSB QSL QSO QSY QTH

Punctuation and procedural characters

$\overline{CT}$	preliminary call
K	invitation to transmit
$\overline{KN}$	specific station to transmit
$\overline{BT}$	double hyphen, =, to separate sentences
BK	break
?	question
/	oblique
8 dits	erase
$\overline{AR}$	end of message
$\overline{VA}$	end of transmission
CL	station closing

See also Tables 3.3 and 3.4 (p4).

The above abbreviations, codes and characters are now used as appropriate in the RSGB slow Morse transmissions (GB2CW) in addition to other material used.

The Novice Amateur Morse Test (5wpm)

In the receiving test the candidate is required to receive a message containing a minimum of 120 letters and seven figures. The duration of the test will be approximately six minutes. Each character will be sent at a speed of 12wpm with a longer-than-normal gap between each character and word to reduce the overall speed to 5wpm. Each character incorrectly received counts as one error. A group of characters, which could include figures in which more than one character is received incorrectly, counts as two errors. More than six errors in this test will count as failure.

In the sending test the candidate has to send, using a straight key, a message of not less than 75 letters and five figures at a speed of not less than 5wpm. This should take approximately three minutes. There must be no uncorrected errors and not more than four corrections will be permitted.

The Amateur Radio Morse Test (12wpm)

The same message length (120 letters and seven figures) is used. In the receiving test this is sent on a straight key in approximately three minutes. More than six uncorrected errors will result in failure.

In the sending test a similar length of message has to be sent in three minutes using a straight key. There must be no uncorrected errors and not more than four corrected errors.

It is not a requirement, of course, that the Novice Test has to be taken before the 12wpm test. However, for some a pass in the Novice Test may be a useful boost to confidence.

General information

A Morse Test application form, list of centres at which tests can be taken and current details of the tests may be obtained from the Radio Society of Great Britain, Lambda House, Cranborne Road, Potters Bar, Herts EN6 3JE.

Your envelope should be clearly marked 'Morse Tests'.

There are many centres at which the tests can be taken, so it is unlikely that you will have to travel very far. Sometimes tests can be taken at the major rallies and exhibitions, but this will depend on the availability of examiners and suitable accommodation.

Fees are currently (1994), Novice Test: £9.00 and Amateur Radio Test: £13.00. The completed application form with the appropriate fee must be sent to RSGB Headquarters.

For identification purposes two passport-sized photographs must be taken to the test.

Disabled candidates who require special consideration must also fill in the Disabled Candidates Application Form which may again be obtained from RSGB HQ. In this case BOTH forms and the test fee must be sent to the Deputy Chief Examiner at the address given on the form. Every possible consideration is given to disabled candidates. In extreme cases a home visit can be arranged.

You may take your own manual (ie straight) Morse key, provided that the connecting leads are terminated in crocodile clips, and/or your own headphones, provided that they are fitted with a standard ¼in mono jack plug.

Do bear the following in mind:

(a) When practising your sending, always remember to correct errors by sending the error signal (eight dits) and then repeat that letter or group.
(b) When receiving, make sure your copy is legible, ie that there can be no doubt what letter or word you have written down.

Those who learn to copy computer-generated Morse may benefit from listening to hand-sent Morse as used in the test.

Don't forget to take a pencil (and a spare) or whatever you prefer to use for the test.

After the receiving test candidates are allowed two minutes to check their copy. Copies are then collected and sent to the RSGB HQ and the Deputy Chief Examiner for processing. Candidates are informed of the result within a few days.

Do not apply for the test until you are pretty confident of success, in other words that you can send and receive at a little more than the required 5 or 12wpm, say, around 7 or 14wpm respectively. Remember that tests are often at rallies at large sites, so allow plenty of time to find the test room. The examiner will be a friendly type who is not there to fail you. There may be another candidate there who is in an absolute panic. Take no notice; say to yourself,"I am all right".

Chapter 8

High-speed Morse and automatic keys

Let us now assume that you have passed the Amateur Morse Test (and the RAE, of course); you have your Class A licence and a unique callsign which is your very own and have been on the air for a year or so.

There are many facets of amateur radio and you may quite likely have chosen one which does not involve telegraphy. Your code is now getting a bit rusty and your speed is dropping. So be it!

On the other hand you may have become fascinated with Morse code as a new language and may have had very many contacts on the key, from the contest rubber-stamp ones to long ones, maybe half an hour to an hour or more with W6 or VK1.

Membership of the 'Rag Chewers Club' requires the submission to the American Radio Relay League of a QSL card which confirms a CW contact which lasted for longer than 30 minutes. The certificate is well worth having.

You are now sending at up to 20-22wpm, but can you take perfect copy at these speeds? In judging your copying speed, do remember that many contacts are stereotyped during the first few minutes; the basic information such as report, location, name etc is often sent at least twice. The chat which follows is sent only once – this is where your copying ability is critical. It is surprising how often the standard of sending drops when it comes to this part of the contact!

As a matter of interest, the American Extra Class licence requires 20wpm, and the Marine Operators First Class licence requires 25wpm. The world record for receiving Morse was set in America in 1939 by T R McElroy at 75.3wpm for three minutes.

Depending upon type, the straight key can be used at up to about 30wpm. Should you endeavour to improve your handwriting at 25wpm? Some marine operators will do copper-plate copy at 40wpm, but not for very long. Generally large amounts of traffic at these speeds will be taken on a typewriter.

It can be argued that it is not necessary in amateur radio to take 100 per cent copy at high speeds, but obviously enough must be taken down or read in your head to enable a sensible 'conversation' to take place. This requires a good clear signal which is not always available in the present-day amateur bands.

Increasing your receiving speed is a case of practice and more practice. The high-speed transmissions mentioned in Chapter 4 will be useful. W1AW, the station of the American Radio Relay League, also puts out high-speed transmissions.

As your speed increases, you will begin to recognise the sound of short words: 'to', 'in', 'of', 'and', 'the'. Also word endings: 'ion', 'ing', 'ley'. This facility becomes very significant as many people find it more and more difficult to distinguish individual Morse characters as the speed increases above about 25wpm.

Now you are beginning to recognise the sound patterns of complete words and you are on the way to becoming a very competent operator.

Many operators tend to develop minor variations in the way they send particular letters or figures. This is generally regarded as acceptable, provided that it is not overdone to the extent that readability is impaired.

Semi-automatic and automatic keys

The semi-automatic key, known universally as the 'bug', originated in America in 1890. The object was to reduce the physical strain of sending for long periods and to increase the speed. No doubt the extra dits in American Morse was another reason for its development.

The bug has two pairs of contacts as shown in Fig 8.1. Dahs are made singly by the separately hinged contact at the front of the horizontal arm. Dits are made automatically by the contact on the rear portion of the arm which is joined to the front portion by a short leaf spring. When the arm is moved sharply to the right, the rear portion vibrates at a frequency which is controlled by the position of the movable weights. Thus dits and dahs are produced by quite different movements of the hand.

How the name 'bug' arose is not clear. A credible story is that the firm which originated the bug (and still makes them) uses the outline of an insect as a trade mark. It would appear that all insects are known as 'bugs' in America, hence the key became known as a 'bug'.

The adjustment of a bug is quite critical. The following notes are abstracted from "Setting up a bug key", H S Chadwick, G8ON, *RSGB Bulletin* August 1951.

High-speed Morse and automatic keys

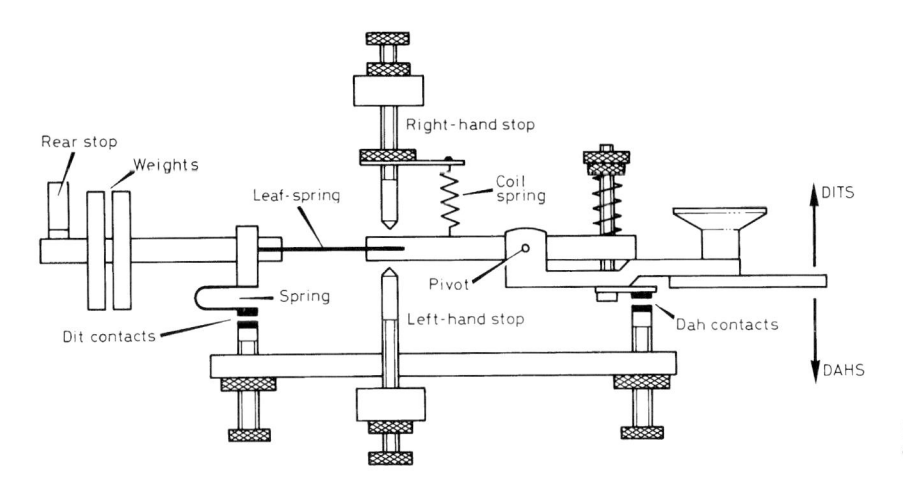

Fig 8.1. Mechanism of semi-automatic key

Initial checks

1. All metallic surfaces which make electrical contact shall be clean.
2. The dit and dah contacts shall be clean and not pitted. If they are not, clean by rubbing the contacts in a perfectly flat position in a drop of metal polish on a plate of glass. Wash the contacts in a solvent and polish with a soft cloth.
3. The horizontal arm shall swing freely with no up-and-down play at the paddle end. If it does not, adjust the bearings at each end of the vertical pivot, and if necessary lubricate sparingly with light oil.

Adjustments

1. Adjust the right-side arm stop so that it just touches the arm when the vibrating end of the arm is just touching its stop. The arm should then be straight.
2. Adjust the left-side arm stop so that there is about 3mm between the vibrating arm and its stop when the paddle is moved to the right.
3. The dit and the dah contacts should be in exact alignment when they are closed.

Fig 8.2. McElroy bug key (circa 1937)

4. Connect a battery, a meter and a suitable resistor in series across the terminals of the key, close the dit contact and adjust the value of the resistor so that the deflection of the meter is almost full scale; note this reading. Send a train of dits lasting up to about 10 seconds or so and observe the meter reading; it should be constant and then start to fall as the vibrating arm slows down. In the period before it starts to decrease, the reading should be about half of the first reading. If it is not, very carefully adjust the spacing between the dit contacts until it is so.

If possible, get a report from a nearby station to confirm that your keyed signal is 'clean'.

The ease and stability of the above adjustments depend upon the mechanical precision of the particular key. The cleanliness and alignment of the dit contacts is especially important. From your experience with a straight key you will be able to set the gap and spring tension of the dah contacts and to appreciate the subtleties of using a bug.

Automatic keys

The el-bug

This is a fully automatic key in which trains of dits and dahs in the correct length ratio are generated electronically. There are two parts to the el-bug: the electronic timing circuits etc, known as the 'electronic keyer', and a single paddle key with a left-right movement. This has separate pairs of contacts on each side so that movement to the right produces a train of dits and movement to the left produces a train of dahs.

The arrangement of this key is similar to the 'sideswiper' which was often home-made many years ago from a piece of hacksaw blade and a few nuts and bolts but which never seemed to be made commercially in this country.

The electronic timing circuits are housed in a small box. They typically include a speed control and an audio oscillator with volume and tone controls for monitoring purposes.

26 The Morse Code for Radio Amateurs

Fig 8.3. A modern bug key (Vibroplex)

PHOTO: JOHN HALL, G3KVA

A separate key may be used, but some commercial versions incorporate the key, with just the paddle protruding through the front panel of the keyer.

The iambic key

This is a more complex electronic key which produces trains of di-dah-di-dah-. . . and dah-di-dah-di-. . . according to which paddle of the twin-paddle key required is operated first.

The paddles are close together and are operated by squeezing between thumb and first finger (hence the often used name 'squeeze' keyer). This key can also be used with the el-bug keyer.

These automatic keys can be used at much higher speeds with much less fatigue. In the hands of the expert they will produce well-nigh perfect Morse.

It cannot be emphasised too strongly that you should not use an automatic key on the air until you have thoroughly mastered the art of sending with it.

You will hear many signals on the amateur bands which are far from easy to read because the sender cannot use an automatic key properly. This is shown by the number of 'spare' dits, in particular, and 'spare' dahs which are transmitted. In fact, a lot of signals originating from a straight key would be easier to read if sent just a little slower with more attention paid to spacing.

Sending speed

There are two golden rules:

1. *Never send quicker than you can receive.*
2. *Never send quicker than the person with whom you are in contact.*

The 'weighting' of a Morse signal

The ratio of the lengths of the dit and dah and the spacing between them was fixed when the Morse code was originated. Some operators consider that at high speeds and in bad interference a Morse signal is easier to read if the lengths of the dit and dah are increased slightly, ie if the 'weighting' is increased. Such a slight increase is obviously difficult to maintain in manual keying.

The more complex electronic keyers are likely to incorporate a weighting control.

This chapter would not be complete without mention of two relatively recent developments: the automatic code reader (Morse decoder) and the CW keyboard.

Fig 8.4. Single paddle key   PHOTO: G4ZPY PADDLE KEYS INTERNATIONAL

Fig 8.5. Electronic keyer   PHOTO: R A KENT (ENGINEERS)

Fig 8.6. Twin paddle key   PHOTO: G4ZPY PADDLE KEYS INTERNATIONAL

Automatic code reader

When connected to the audio output of a receiver, the code recorder reads the incoming Morse transmission and displays the Morse characters as scrolling letters and figures on, typically, a 16-character liquid crystal display.

The particular instrument illustrated in Fig 8.7 also reads radio teletype (RTTY). In addition, it supplies two tutorial facilities, ie generation of groups of random Morse characters, and it will display Morse sent to it so that spacing between words and letters can be checked.

The less-complex readers are not necessarily effective on very weak signals or signals in bad interference.

The CW keyboard

This is similar in appearance to a computer keyboard, but has additional keys for Morse procedural signals such as $\overline{AR}$, $\overline{AS}$, $\overline{VA}$ etc. It will produce perfect Morse at whichever speed you are able to type. Normally it has the facility for storing the message for transmission at a lower preset speed. Such devices may be an alternative for those who are unable to use a conventional key.

No comment is made on the use of a Morse decoder in conjunction with a CW keyboard which enables one to send perfect Morse at any speed and to read an incoming Morse signal on a visual display without any knowledge of the code. Many Morse operators may regard them with horror. Others may use them, giving a false impression of their competence.

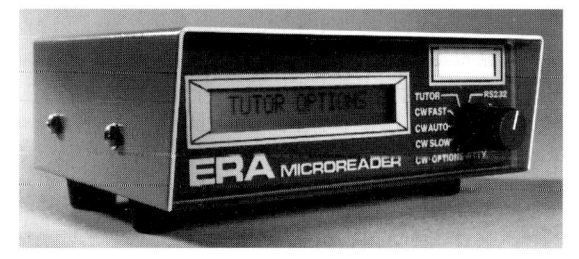

Fig 8.7. Automatic code reader   PHOTO: ENTERPRISE RADIO APPLICATIONS

Further reading

The *Handbook for Radio Operators* published by Her Majesty's Stationery Office is the manual for marine radio operators. It does not contain much of direct use in amateur radio but is worth a quick read. It is suggested that you borrow a copy from your local library as it is now fairly expensive.

The Morse code enthusiast will find much of interest in the quarterly journal *Morsum Magnificat* which is now published in this country by:

> G C Arnold Partners
> 9 Wetherby Close
> Broadstone
> Dorset BH18 8JB

A final note

You have now completed your study of a quite comprehensive book on the Morse code as practised in amateur radio.

You are probably now competent at 30/35wpm; maybe you are wondering about the HSC (High Speed Club) or the VHSC (Very High Speed Club), both of which are based on the Continent. Membership of these select organisations requires nomination by several members with whom you have had contacts at 40wpm or more. If you do not bother to achieve these dizzy heights, it does not matter. Your enjoyment of Morse telegraphy will be just the same!

One final comment: please do not forget the second golden rule – "Never send quicker than the person with whom you are in contact."

Appendix

Miscellaneous Morse characters

1. Abbreviated figures

1	di-dah	6	dah-di-di-di-dit
2	di-di-dah	7	dah-di-di-dit
3	di-di-di-dah	8	dah-di-dit
4	di-di-di-di-dah	9	dah-dit
5	di-di-di-di-dit	0	dah

In commercial practice these may be used *provided that no misunderstanding can arise from their use*.

In amateur radio the standard signal report given in contests (irrespective of signal strength!) is 599, sent as 5NN. Serial numbers sent in contests are quite often given as a mixture of abbreviated and standard figures. This does cause confusion, which results in a request for a repeat.

0 (zero) is often sent as a double length dah, but *never* in a callsign. Dah-dah-dah should never be used for 0.

2. Accented letters

These are only heard occasionally on the amateur bands. They are included for reference only; do not bother to memorise them.

ä	di-dah-di-dah
á or å	di-dah-dah-di-dah
ch	dah-dah-dah-dah
é	di-di-dah-di-dit
ñ	dah-dah-di-dah-dah
ö	dah-dah-dah-dit
ü	di-di-dah-dah